Collective Bargaining

A Canadian Simulation

Collective Bargaining

A Canadian Simulation

W. Paul Albright

School of Business and Economics
Waterloo Lutheran University
Waterloo, Ontario

Wiley Publishers of Canada Limited

Toronto

Copyright © 1973 by Wiley Publishers of Canada Limited
Published outside Canada by John Wiley & Sons, Inc.
(New York, London, Sydney)

No part of this book may be reproduced by any means, nor transmitted, nor translated into a machine language without the written permission of the publisher.

Library of Congress Catalog Card No. 73-5092
ISBN 0-471-02049-4

All photographs, with the exception of that used on page 1, are from Miller Services Limited, Toronto.

Cover design by K B Graphics.

Printed and bound in Canada

Contents

	Preface	*vii*
1	Game Description	*1*
2	History of Duro Metal Products, Limited	*15*
3	Duro's First Collective Agreement	*25*
4	Background Notes on Duro's Present Collective Agreement	*31*
5	The Present Agreement	*43*
6	Costing the Package	*57*
7	Three Selected Issues	*61*

Appendices

I	Contract Ratification by Union Members	*71*
II	Divergent Problem Solving and Collective Bargaining	*73*

III	Industrial Relations in Canada Today—An Overview	75
IV	The Office Furniture Industry in Canada	83
V	Team Strategy Report	87
	Final Negotiation Report	89

Preface

This manual describes the procedures for a simulation in which students, acting as management or union representatives, bargain a complete labour agreement for a mythical Canadian company, Duro Metal Products, Limited. It is designed to provide negotiating experience for the players, and to help them develop the personal skills mandatory for them to be effective in task-oriented groups which work under a certain amount of pressure.

My experience at the bargaining table, both as a spokesman for management and as a member of several arbitration boards, has made it possible for this simulation to closely approximate real collective bargaining.

The game has two important characteristics. Firstly, it runs itself: students using the manual require no supervision. Secondly, and more importantly, the outcomes of the different teams' bargaining can be scored objectively by an instructor with no collective bargaining experience whatsoever: complete scoring instructions are provided.

Anywhere from eight to twelve hours of in-class time and eight to twelve hours of out-of-class time will be required, depending on the time available and the depth of treatment desired. No specialized prior knowledge is necessary, because the format is simple and easily understood; the level of sophistication will be a function of the degree of expertise and maturity that the players bring to the bargaining table.

This simulation has been successfully class-tested on some one thousand students in Industrial Relations, Human Relations, and Introductory Business courses, among others. It was found that the participants became personally and deeply involved, and that many were able to discover some interesting things about how they, and their colleagues, react under stress.

Background material, provided to frame the bargaining simulation in a realistic industrial and legal setting, is found in the appendices. When the simulation is used as part of an Industrial Relations course, additional readings can be assigned.

To conclude these opening remarks, I would like to acknowledge the help of several colleagues—particularly that of Dr. John Weir—and special thanks to Wiley's Production Editor, Anne Rutherford Sinclair.

W. Paul Albright
Waterloo, Ontario
April 1973

1
Game Description

This simulation is a collective bargaining experience. You will be a member of either a management or a union team and will bargain with your opponents in a series of meetings in an effort to arrive at a contract for the hourly rated employees of Duro Metal Products, Limited, a company located in Calgary, Alberta.[1]

The purpose of this game is to:

(a) develop your negotiating skills;
(b) increase your awareness of the importance of understanding (but not necessarily agreeing with) the point of view of others whose objectives differ from your own;
(c) help you to understand the forces which shape collective bargaining.

There will be several bargaining groups, each consisting of a management and a union team. The members of the teams are:

Representing the Company

Director of Industrial Relations
Production Manager
Assistant to the Vice President: Marketing
Comptroller

---

1. Neither Duro Metal Products, Limited, nor the union—the United Fabricators of America—actually exists. Both, however, are realistic composites of actual companies and unions in the metal fabricating industry.

Representing the Union
International Representative
President: Local Union
Vice President: Local Union
Committeeman: Press Shop
Committeeman: Assembly Department

You will work with your teammates and bargain to meet several somewhat conflicting objectives simultaneously. These objectives include:

(a) helping your team to bargain the best possible contract;
(b) getting as many as possible of the particular benefits most wanted by you as an individual;
(c) helping to avoid a strike, unless it develops that this is the least unattractive of the undesirable alternatives finally facing you.

Successful bargaining requires teamwork and collective decision making. Your influence on the outcome will depend, in large part, on the respect which others have for your competence, knowledge, and judgment. Ultimately in a bargaining situation, authority does not derive primarily from rank or position; it must be earned. It will probably be closely related to the extent to which others interpret your actions as being helpful to them in furthering their own particular objectives.

EVALUATION OF PERFORMANCE

When the game is over, the instructor will compute a score for each player, based primarily on results. Company representatives will be given merit points; union representatives, political points.

Your final individual score will be based on two factors, the most important being your team's success in bargaining a good overall contract. The other criterion will be the extent to which the final contract places emphasis on the particular benefits which you as an individual value highly.

Management-Team Performance

Merit points for management teams, at least over a fairly wide range, are inversely related to the increased costs imposed on the company by the changes in the contract. There is, however, an exception: if the cost of the final contract or final offer is unduly low, whether due to the union team's incompetence, bad luck, or whatever, the company team will receive a low score.[2] A union team, bargaining such a contract, would be thrown out of office in short order.[3] Further, union or no union, these people still work for Duro, and reasonably competitive working conditions are necessary if Duro is to get and keep good employees.[4]

The relation between management's merit points and the cost of added benefits is shown in figure 1.1. You will note the absence of units on the axes; part of your job, of course, is to estimate these as bargaining progresses.

It is important for management teams to note that a firm can lose its competitive position in an industry by careless bargaining. In the first place, if a company's labour costs are too high, its profit position is eroded and it is gradually starved of the funds needed to compete effectively and to grow. In addition, if, in an ill-advised and short-sighted attempt to avoid a strike at all costs, a firm's representatives bargain a succession of contract clauses which prevent them from managing their plant effectively, they may find they have been "nibbled to death". Firmness in bargaining, even if this means an occasional strike, is better than loss of control of the plant.

2. V.E. Vroom, *Work and Motivation* (New York: Wiley, 1964).

3. For the purposes of this game, assume that both management and union teams have full authority and that the deal which they make is final. In practice, of course, there is often another step for the union—ratification by the general membership (see appendix I).

4. The relationship between the cost of added benefits and the management-union points shown in figures 1.1 and 1.2 applies to Duro, but not necessarily to all bargaining situations. For instance, a company might decide to try to get rid of the union if it thought conditions were right (weak union, rash union bargaining committee, substantial employee discontent with the union, etc.). Under such conditions a company might elect the strategy of making and holding to a very low position, hoping to goad the union into rash action.

FIGURE 1.1 Relation between Cost of Added Benefits and Management-Team Merit Points

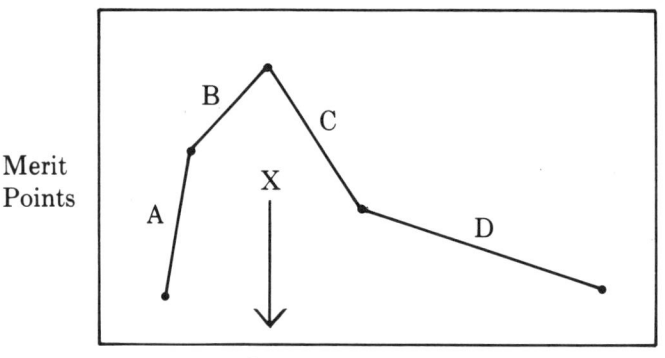

Region A: Contract ridiculously low.
B: Added benefits not enough to compete properly in labour market.
C: Added costs jeopardizing company growth rate.
D: High C costs jeopardizing solvency should a recession occur.

Union-Team Performance

Generally speaking, political points for union teams are also closely related to the total cost of the changes bargained. Over a fairly wide range, the more costly to the company, the more points for the union teams. If the contract is excessively costly, however, the union teams earn few political points because additional factors come into play. Benefits beyond the point required to guarantee re-election of the union's officers become increasingly offset by two negative factors. First, high labour costs may lead to excessive layoffs; and second, "too good" a contract at Duro could be a source of embarrassment for the union when it bargains for workers in other companies which cannot or will not meet similar terms. The relation between political points for union teams and contract costs is shown in figure 1.2. Again, there are no units on the axes; you must estimate appropriate ones.

FIGURE 1.2 Relation between Cost of Added Benefits and Union Team's Political Points

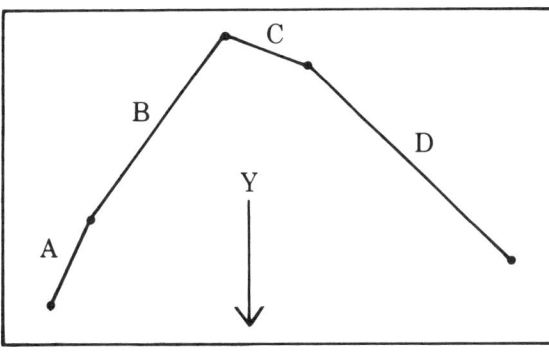

Political Points

Cost of Added Benefits

Region A: Benefits low, employees dissatisfied, high probability they will switch to a different union.[5]
B: Re-election of present union officers jeopardized.
C: Excessive layoffs more likely.
D: Duro contract an embarrassment to the union when bargaining elsewhere.

There are two possible bargaining outcomes for each group. Either management and union agree on a contract, and there is no strike, or they do not agree, and there is a strike.

If there is agreement, teams will be assigned points in the manner shown in figures 1.1 and 1.2 and there will be no strike penalty.

If there is no agreement, both sides will be assessed a strike penalty as well as a further penalty, the size of which will depend on how far their *final* offer or demands depart from a realistic position.

---

5. When a contract expires, if employees are disenchanted with their union it can be decertified. There will then be no union, or, if the employees wish, they can get a different one.

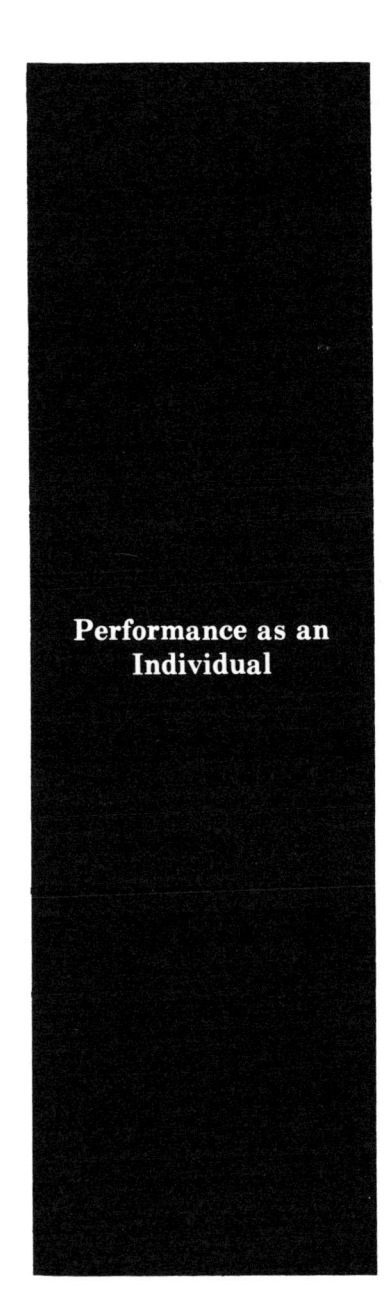

These positions, shown as points X and Y (union and management), are known to the instructor.

Though you will be aware, no doubt, of strong opposition from across the table, your real competition comes from teams on your side in the other groups. Management's performance will be evaluated by comparison with other *management* teams, and similarly for union teams.

In a team effort such as this, some individuals usually make a greater contribution than others. This factor should be recognized but is difficult for an outside observer to evaluate when many groups bargain simultaneously. Peer rating is one way to overcome this problem; your instructor may ask you to rate the contribution of each of your teammates. Such a procedure will allow him to adjust the team score of individuals on each team slightly up or down from the team average, which he will determine.

Performance as an Individual

Once your role is established, you will be provided with a Personal Profile. It describes your background, beliefs, priorities, and objectives. Your profile will differ considerably from those of the people sitting on the opposite side of the table and will also differ in some respects from those of your teammates. For instance, some union representatives may prefer a modest increase in the hourly rate plus increases in the pension plan as opposed to a larger hourly rate increase with no pension increase, even though the costs of the packages are identical. Older employees who value security may want retraining and transfer rights because they fear that their job may disappear through technological change. Younger employees, possibly more mobile and more willing to take a chance, may prefer better promotion opportunities. The priorities of individual company representatives also differ. Some place a high premium on uninterrupted production and sales and consequently are more anxious than some of their colleagues to avoid a strike. Some prefer certain benefits because they do not readily apply to nonunion employees. If, for instance, the union bargained extra paid holidays, office employees would expect similar treatment. Additional layoff benefits, on the other hand, might well not apply in the office.

RULES OF THE GAME

Your success in obtaining the personal objectives implicit in your profile will be evaluated by comparing your performance with that of your own counterpart on other teams.

You must bargain in good faith. This does not mean that you have to compromise and shift your position to the point where your vital interests are seriously jeopardized. It does mean that you must accept the opposing side as bargaining agents with the authority both to make and live up to an agreement. Further, the opposite side must get a fair hearing and have a chance to develop its position. You cannot, for instance, take a stand, and then issue an ultimatum to the effect that if they do not "immediately do this and so", you will attend no further meetings. This of course does not preclude either side from calling for a caucus, so that they can sort things out in private.

In playing the game you have considerable leeway as to sources of information. You may want to refer to outside readings and to contact knowledgeable people in industry for general background material. If you use outside information it must not conflict with information supplied in the player's manual. As the game progresses, an important source of information will be the team on the opposite side—its arguments, statements, actions and reactions—though no doubt you will want, appropriately, to discount these as bias, bluffs, and attempts to gain tactical advantage.

You may feel at times that the manual does not give enough information. This reaction is quite normal and is also typical of actual bargaining in industry. In any event, you have all of the information that there is, and your opposite numbers in other groups have the same problem. As a matter of fact, one of the objectives of this simulation is to help you to develop the art of making decisions when you have only partial information and are under the pressure of time (see appendix II).

One source of information is forbidden. It is against the rules to discuss the game in any detail whatsoever with team members of the opposite side in your own negotiating group or with either side of other negotiating groups.

With the exception that each member must play the role assigned, you may organize your team in any way that you wish. You may or may not have one or more spokesmen, delegate authority and responsibility as you please, have team members prepare written reports for team meetings, and have team meetings outside of formally scheduled class sessions. You should select a chairman. Formal bargaining sessions will be at the times and of the duration described later in the manual. Each bargaining group may work out whatever rules of procedure are necessary to keep some degree of order in the formal bargaining sessions.

You are not obliged to tell your opponents the *whole* truth. You should, however, keep in mind that to be caught outright in a direct lie will diminish your credibility, increase the uncertainty, and make it more difficult to arrive at a sensible contract.[6] Both sides must present their best arguments and you need not feel compelled to do your opponents' bargaining for them. Some of your interests will coincide but others will not. You will probably find that sweet reasonableness will not resolve all conflict; ultimately, *power* (the ability to influence the likelihood of others obtaining *their* objectives) will be a factor.

You will find it advantageous to plan ahead so that you can deal with the responses and arguments of the opposing team effectively.

SCHEDULE FOR MEETINGS AND TASKS

Session 1: Organizing

(60-120 minutes)

(a) Meet as a class for instructor's introduction of game and divide into groups of nine—four company, five union.

---

[6] This is also important in a real bargaining situation because usually a continuous relationship exists; the opposing parties must: (a) administer the contract bargained; and (b) negotiate a new one when the present one expires. When they work this closely together, if either side has told obvious lies they are usually found out—and their credibility suffers.

(b) Separate into teams and:

- (i) each team member select his role;
- (ii) each team select a chairman;
- (iii) organize teams: assign individual's responsibility and authority, and work assignments; decide on out-of-class sessions, if any;
- (iv) begin development of team strategy;
- (v) delegate responsibility for preparing the Strategy Report (see session 2).

* Time Interval

Session 2: First Team Strategy Session

(90-120 minutes)

(a) Establish initial bargaining position: list all changes involved and price the total package (in cents/hour).

(b) Establish in general terms your intermediate (for session 5) and final (for session 7) positions. Specific contract items need not be itemized here but the total cost of the planned intermediate and final positions must be computed.

(c) Decide on the content of a Team Strategy Report. This is to be given to the instructor before the beginning of session 3. This report will list in detail the decisions arrived at in (a) and (b) directly above (see appendix V).

* * * * * * * * * * * * * * * * * * * Time Interval

Session 3: First Bargaining Session

(60-75 minutes)

(a) Both teams outline their position (union first) in full detail and with succinct supporting arguments. Time allowed: 20 to 40 minutes per team.

(b) After both presentations are completed, union and company teams will react to the opposing side's proposals, commenting on the implications.

* * * * * * * * * * * * * * * * * * * Time Interval

Session 4: Second Team Strategy Session

(60-120 minutes)

Review strategy in light of the first bargaining session. Once bargaining begins you are not compelled to hold to the positions spelled out in your strategy report; these were tentative plans. When revising your strategy take into account your estimate of the priorities and acceptable minimum of the other side.

* * * * * * * * * * * * * * * * * * * Time Interval

Session 5: Second Bargaining Session

(60-120 minutes)

Continue bargaining, either shifting positions and giving arguments, or holding old positions with reinforced arguments.

* * * * * * * * * * * * * * * * * * * Time Interval

Session 6: Final Strategy Session

(60-90 minutes)

(a) Decide on your tactics for the final bargaining session. Establish a tentative position beyond which you will not go even if a strike results.

(b) Delegate a team member to prepare the team's Final Negotiation Report listing in detail (with costs) the terms of the final contract (if agreement reached); or, a detailed list (costed) of your team's final position (if agreement not reached).

* * * * * * * * * * * * * * * * * * * Time Interval

Session 7: Final Bargaining Session

(60-150 minutes)

(a) Conclude bargaining.

(b) Be certain there are no misunderstandings about either the terms or any important wording and have both team chairmen initial a draft of the other team's final negotiation report. At the end of the session, or shortly thereafter, as instructed, submit this report to the instructor (see appendix V).

ADDITIONAL REPORTS

Each *side* is to prepare a short report to the instructor, with its members evaluating their team's successes and failures in both strategy and tactics, and noting particularly what, if anything, they would do differently if they had it all to do again.

Each individual player is to prepare a 300-word report to the instructor describing what he learned from the game. This should include an opinion of the simulation as a learning experience, rated: excellent, good, fair, poor, or very poor. Do not sign this report.

2
History of Duro Metal Products, Limited

Duro Metal Products, Limited, a medium-sized company located in Calgary, Alberta, is primarily engaged in the manufacture of metal office furniture. It also produces a line of office components, including wall cabinets and drawers, work surfaces, office-machine surfaces, and wall panels, which used together in various combinations provide the modules for "office landscaping" systems. In addition, Duro manufactures a number of other stamped metal products, such as steel shelving, garage doors, and lockers.

The company owns approximately eight acres of property in the older industrial section of the city. On this property are located the plant and offices—which are housed in a fifty-year-old, two-storey building—warehouse facilities, employee parking, and a spur line providing railway service.

The firm was founded in 1962 by two brothers, William and Benjamin Armstrong. William had worked as a sheet metal worker for several years and had been a foreman in a metal fabricating plant. Benjamin had twenty years' sales experience and for several of these years had sold office furniture for a large company. Confident of the growth potential of the West, both believed that there was a great opportunity for a new company selling high-quality office furniture of modern design. They were convinced that they could develop more efficient manufacturing methods than were typical of the industry, and that, protected by the Canadian freight rate structure, they could, within a few years, serve a region stretching

Company History

from the Lakehead in the east to the Pacific Ocean in the west. They formed a partnership and with $60,000 of their own money, four employees, and some used machinery they started operations in a small, rented building.

Benjamin was responsible for marketing; William looked after manufacturing and administration. Benjamin worked on building a reputation for high quality, good design, dependable delivery, and realistic (but not low) prices. William was a good organizer and strictly controlled factory operations. He soon made it clear who was boss in the plant, expecting and getting high performance from his employees. Believing that once policies, objectives, and delivery schedules were made clear, most employees could accept responsibility, he allowed considerable leeway for them to work out the best methods and timing for individual orders. It was not unusual to see plant workers discussing product design with the product designer and the machine tool designer. Both brothers drove themselves hard and worked long hours. They believed in spending money to make money, hiring competent employees and paying somewhat better-than-average wages and salaries. Further, individual employees whose work was considered outstanding were given merit raises which were over and above their basic earnings.

By 1967, results had fully confirmed the partners' original convictions. They produced quality products on schedule at low cost and were able to grasp quickly the multiplying market opportunities. Problems, however, were beginning to develop.

The company was not growing as quickly as they wished. More space and new machines were needed for maximum efficiency. In addition, they were turning away many orders, to the point that they were afraid some competitor would establish a secure foothold in markets which they could serve.

The partners decided to expand by going public and in May 1968 the firm was reorganized. At this time the present property and additional equipment were purchased. William became president and Benjamin, vice president—Marketing. A production manager, sales manager, and comptroller were appointed.

Now that William was more involved with new equipment purchases and finances, and Benjamin with special orders for large office and institutional buildings—a type of business they had previously been unable to handle—both men found it more difficult to keep in touch with day-to-day operations. They had full confidence, however, in the managers whom they had selected and gave them a free hand. As Benjamin said: "There is no use in hiring hounds and then doing all the barking."

By the spring of 1969 they were settled into their new location. They also faced new challenges. Sales, though increasing, were short of forecasts, and though the books showed good profits, there were often rather serious cash problems. In addition, trouble seemed to be brewing in the plant. The new production manager and the comptroller were both very cost conscious and they exercised their considerable authority in ways which often upset the employees. Frequently, for instance, parts which were formerly made in the plant were contracted out to other firms. Sometimes this was done so that the company could accept orders which otherwise would have been turned away; other times it was done because it was cheaper. This practice had resulted in some employees' having to accept transfers to undesirable jobs. In one instance, when machinery made their skills redundant, two employees quit rather than accept the only jobs available. In that year most of the employees were angered when the company granted smaller-than-expected annual wage increases.

The easy team spirit which had existed in the early days was now gone but output continued to be high because of close supervision and pressure. Detailed production schedules were instituted and a time-study man hired to establish standards for most of the jobs. Management knew that the men were upset but believed that they would appreciate that the company's first priority was to grow quickly and become Number One in western Canada before a competitor beat them to it.

It was soon apparent that the employees did not see things this way. In fact, many felt that success had gone to William's head and that his ambition had turned him against them. The United

Company History

Fabricators of America, already active in the region, ran an organizing campaign at Duro, and with the active help of a particularly disgruntled employee, signed up three-quarters of Duro's factory employees in a matter of days. In spite of company efforts to prevent it, in November 1969 the Fabricators were certified as the exclusive bargaining agent for the hourly rated employees (see appendix III: "Industrial Relations in Canada Today—An Overview").

The company was shocked by the apparent ease with which the union organized the plant. As William put it at the time: "After all we've done for them you wouldn't think some glib-tongued outsider could con them into joining a union." Management finally had to sit down at the bargaining table and a one-year contract, expiring in February 1971, was negotiated with the union. Subsequently another contract was signed; this ran from March 1971 to December 1972. (A more detailed account of labour-management relations during these years as well as the full content of the agreement signed in March 1971 appears in the next chapter.)

During the period 1969 to 1972 the company continued to grow steadily; Duro served its customers well and gave them quality, modern design, realistic prices, and, notwithstanding sporadic but minor labour troubles, dependable delivery. William attributed much of this success—the low production costs and the high profits—to the introduction of an incentive plan which had been negotiated as part of the last contract.

An Income Statement and Balance Sheet for 1972 plus a tabulation of income and sales for the period 1965 to 1972 follow. Additional data on the industry and the economy appear in appendix IV.

TABLE 2.1   Duro Metal Products, Limited
Balance Sheet: 31 December 1972

Assets	
Cash	$ 80,000
Marketable Securities	200,000
Accounts Receivable	1,120,000

TABLE 2.1 (Cont'd)

Inventory		
Material, Supplies	$640,000	
Goods-in-Process	184,000	
Finished Goods	300,000	
Purchased for Resale	216,000	
Total Inventory		$1,340,000
Prepayments		40,000
Total Current		$2,780,000
Net Buildings and Equipment		1,200,000
Total Assets		$3,980,000

Liabilities

Accounts Payable	$ 940,000
Notes Payable	600,000
Total Current	$1,540,000
Mortgage (secured by plant)	560,000
Total Liabilities	$2,100,000
Common Stock	$800,000
Retained Earnings	1,080,000
Total Liability and Shareholder Equity	$3,980,000

TABLE 2.2 Duro Metal Products, Limited
Income Statement for the Calendar Year 1972

Net Sales		$6,400,000
Cost of Goods Sold		
Materials and Supplies	$3,400,000	
Production and Related		
Salaries and Wages	950,000	
Other Production Related	690,000	
Total Cost of Goods Sold		$5,040,000

TABLE 2.2 (Cont'd)

Gross Margin		$1,360,000
Administration and Office	$400,000	
Sales and Distribution	260,000	
Total		$660,000
Operating Income		$ 700,000
Other Deductions (interest, etc.)		120,000
Net Income Before Tax		$ 580,000
Income Tax, Net		286,000
Net Income		$ 294,000

TABLE 2.3   Duro Metal Products, Limited
Net Sales and Income, 1965-1972

Year	Net Sales	Net Income
1972	$6,400,000	$294,000
1971	5,578,000	202,000
1970	4,596,000	269,600
1969	3,202,000	170,000
1968	2,006,000	89,800
1967	854,000	61,000
1966	682,000	49,400
1965	486,000	39,800

TABLE 2.4   Number of Hourly Rated Employees
at Duro by Job Classification

Job Classification	Hourly Rate	Number
1	$2.90	19
2	3.00	25
3	3.15	15

TABLE 2.4 (Cont'd)

Job Classification	Hourly Rate	Number
4	$3.30	10
5	3.45	10
6	3.60	8
7	3.75	8
8	3.95	5
9	4.20	4

Average Hourly Rate = $3.27 (excluding incentive*).

*Approximately 60 per cent of the above jobs earn incentive pay over and above these basic rates. The average employee's incentive pay is 20 per cent over his base rate.

3
Duro's First Collective Agreement

Duro's management was angry and worried when the employees turned to the union. William, especially, was convinced that the union would try to restrict the company's freedom in many ways, such as the right to contract out work to other firms. He was also afraid productivity would suffer if the union pressed for seniority as the criterion for layoffs and promotions; he believed this practice destroyed personal initiative.

Most of all, he was concerned about costs and what he considered the "Alice-in-Wonderland" wage increases being bargained by some unions. Low costs were still vital to Duro's profit and growth. Though their office furniture and the office landscaping component lines were both marketed under the Duro brand name and had an image which protected them from cutthroat price competition, nevertheless, as Benjamin continually emphasized, the industry was competitive and big buyers were cost-conscious. Further, the market for other items such as lockers and shelving, products which the company made primarily to keep the plant running to capacity when other product lines were slack, was extremely price-sensitive because of the many small marginal firms in this field. In fact, the prices on such items barely covered variable costs.

When the Fabricators applied to the Labour Relations Board for certification, the company acted at once. They fired the assistant to the production manager for not keeping them better informed about the deteriorating plant morale and about union organizing moves.

They also released—allegedly for incompetence—an employee who was known to have helped the union. Finally, they obtained the services of a management consultant who specialized in labour relations. His first job was to prevent certification if possible, because they did not believe that the majority of the employees supported the union; they felt that the "night riders"[1] must have used high-pressure tactics. He was also to help with negotiations if that became necessary.

At the certification hearing the company presented a counterpetition signed by over half of the employees. This petition said that they did not want a union. The union, on its part, asked the board to examine the circumstances surrounding the discharge of the union worker; they claimed it was a clear case of discrimination for union activity. The board ruled against the company on both counts, stating:

(a) that there was good reason to believe that the company had applied pressure in getting the signatures on the counterpetition; and
(b) that there was not enough evidence of incompetence to warrant discharge of the employee.

The board then ordered him reinstated.

The negotiations for the first contract were conducted in an atmosphere of hostility. The company team was led by the labour consultant, who was acting under instructions to drag things out and make it as difficult for the union as possible. The company hoped that the employees would "come to their senses" if they saw the union making no progress. This approach, however, only succeeded in alienating the employees further. Seeing no progress in the first two months of negotiations, they began to take matters into their own hands: production was delayed for several days when large numbers of key employees reported "sick", and productivity dropped when some employees took part in what appeared to be an organized

---

1. A derogatory term sometimes used by management when referring to union organizers who visit employees' homes in the evening to try to persuade them to sign union cards.

slowdown. Finally, there were attempts to block a train which was shunting in a carload of steel. During this episode William was heartily booed by the men as he stood on the roof of the warehouse, taking pictures with his movie camera when men were lying across the tracks.

Faced with rising costs and upset production schedules, management finally decided to negotiate in earnest. By this time the labour consultant had convinced them that there was a pervasive sense of grievance in the plant and that they would have to come to terms with the union. Management was determined, however, to stay in control of operations and to keep costs down. The union could see that they were deadly serious. A contract was finally negotiated which gave the employees only a small wage increase and did little more than put into writing the employee benefits which were already in effect. The company fought hard against seniority and insisted on a strong management's-rights clause which would give them complete freedom to make or buy parts pretty well as they saw fit. They also resisted a strong demand for a union shop, which would have established that all employees must be union members as a condition of continued employment.

The union, naturally, did not like the contract but felt it was the best that they could do without dragging things out interminably, something they did not want to do because the employees were getting restless, and there were rumours that another union had approached some of them, suggesting a switch if the Fabricators could not get a contract.

The bargaining committee outlined the terms of the contract to a general membership meeting in a local hall. The younger members of the union in particular did not like the small wage increase and spoke out strongly against accepting it. The committee, however, convinced the members that it was a good start and the meeting voted to accept it—65 per cent in favour, 35 per cent against.

4
Background Notes on Duro's Present Collective Agreement

In Canada and the United States, the collective agreement is the cornerstone of collective bargaining.[1] When a company and a union agree on the terms of employment they draw up and sign a contract which holds for a specified period of time, usually one to three years. This agreement is like legislation; it is the basis for, and governs, the relations between the union (which speaks for the employees) and management. Like much legislation, however, many clauses are written in general terms which simply express agreement in principle. The total of the day-to-day relationships of a company with its employees would require several large volumes if spelled out in enough detail to cover every situation, and even then many new situations not envisaged would arise. Actually the agreement is the skeleton of the relation between management and the union. In practice, this skeleton is fleshed out by memos, bulletins, various plant rules, custom, and oral understandings.

Collective agreements differ considerably. They differ in length: some have two hundred pages, others only twenty. The content

1. The situation is different in many countries. During a visit to Ireland a few years ago the writer asked an industrialist for a copy of the agreement covering his plant employees, only to be told that they did not write such things down in one place! This informality contrasts sharply with Canadian practice. Here the agreement is printed in a pocket-size booklet for employees, foremen, and stewards. The union stewards, especially, use it as their bible; within a few weeks of printing, their copies are well thumbed and dog-eared.

differs too, depending on the technology of the industry. If layoffs and cutbacks, for instance, are common (automotive), there will be detailed clauses specifying which employees go and which employees get the available work. If such problems rarely come up (in white-collar unions, for example), one short paragraph often suffices. Agreements also differ depending on the duration and the nature of the relationship between a company and a union. Contracts tend to become longer over time, and issues which have been the object of disputes tend to be spelled out in greater detail to avoid misunderstandings in future.

Though the wording will often differ drastically, some types of clauses appear in almost all agreements—clauses establishing both parties' responsibilities and rights (union-recognition and management's-rights clauses), clauses dealing with wages and fringe benefits, clauses dealing with job security and job opportunities (seniority), and clauses outlining the machinery for settlement of disputes (grievance procedure). You will note that the Duro Agreement which follows has all of these. In addition, most contracts have clauses which are of special importance to the industry or the company—perhaps a detailed outline of how time studies will be conducted (auto), a procedure for employee transfer between stores (supermarkets), job evaluation procedures (steel), or incentives (Duro).

THE IMPORTANCE OF THE GRIEVANCE PROCEDURE

Before considering the clauses and issues in the present agreement which will come up in your bargaining in this game, you should note article 12, beginning on page 52. This is the grievance procedure and it is important for an understanding of the fundamentals of collective bargaining. You will not be bargaining this because it is quite similar in most contracts with industrial unions. This procedure has settled into its present form as the result of decades of experience of management and unions. There are several points worth noting. First, there are rigid time limits built into the various steps. Surprisingly, perhaps, not only the union but also the management has found this advantageous. Smouldering, unresolved grievances are dangerous to both, and justice delayed is often justice denied. Note, too, the "statute of limitations" (section 12.07). This protects the

Background Notes on Present Agreement

company from grievances brought forward long after proper facts can be obtained; in addition, it allows the company to take action and exercise leadership and have most of its acts legitimized by the silence of the union.[2] Note, too, the rapid escalation of grievances to high levels in the company. Companies, in agreeing to these procedures, implicitly recognize the importance of "people problems". Top management has many important problems to deal with all the time. Nevertheless, these must wait until grievances are investigated and decided upon; otherwise the grievance is conceded by default.

The grievance procedure describes the method of settling all those disputes (over the meaning of certain clauses) which may arise during the life of the agreement—who will get promoted, who gets the overtime, etc. Some items, such as number of days' vacation, are clear; others, like "just cause" for discharge, are not so clear, and both parties are disposed to read in their own interpretation. Union officials view the grievance procedure as one of the most important parts of the contract. In effect, every union employee has a judge standing behind him, in that he has free access to objective, impartial arbitration of his grievance (provided the union officials agree that the grievance has merit).

Canadian agreements make provision for compulsory and binding arbitration as the last step in the grievance procedure if all else fails. The company and the union select an arbitrator or sometimes a board of three. Arbitrators are usually judges or sometimes university professors. They hold a hearing, take evidence, and listen to the opposing arguments. The basis for the decision is the contract and this the arbitrator has no power to change; he can only interpret, having in mind the evidence and the most probable intent of the parties when they drafted the agreement. If the actual words are not clear, or if reasonable inferences cannot be drawn from the contract wording, then past practice and precedent will be taken into account as indirect evidence of the intent of the parties. It is very important, therefore, when drafting the wording of contract clauses, to guard against language which is capable of unfavourable interpretation.

2. Here labour relations resemble romance; a real Casanova does not ask for a kiss, or what have you—silence gives consent!

Background Notes on Present Agreement 35

Arbitration has worked well in Canada as a method of enforcing labour contracts and has minimized the frequency and severity of work stoppages and slowdowns.[3] There are certain advantages over the courts: the arbitrators are specialists in labour matters, the cost is much lower, and rulings can be obtained more quickly. Most importantly, fact finding and informal procedures are used rather than strictly legalistic ones. The "balance of probabilities", for instance, is often used as the standard of proof rather than "beyond reasonable doubt", as in criminal court proceedings.

It is important to distinguish between arbitration and conciliation. Arbitration, except in special circumstances, does not apply to disputes over the terms of a new or renewed contract. These are settled by bargaining, with or without a strike or lockout, and often with the involvement of a government-appointed conciliator whose job it is to help the parties come to an agreement. Recommendations of a conciliator or of a conciliation board are rarely binding on the parties.

NOTES ON SPECIFIC CLAUSES IN THE PRESENT AGREEMENT

As you read over these comments on the present contract, keep in mind that there are three articles: Recognition (article 2), Management Functions (article 6), and Seniority (article 10, section 8) which you bargain by selecting one of several clauses already written for you. Any changes bargained in the other clauses should be written out and initialled by both team chairmen. Either side may bring changes to the bargaining table, and may also eliminate or add articles or sections.

Article 2: Recognition

You will note that the recognition clause now in the Duro agreement affords little security to the union; the company simply agrees to collect union dues by payroll deduction, and employees can join or quit the union at will. The union found it difficult to get even this

---

3. Arbitration is less favoured by skilled-trades unions. Their members often work for several employers in a short period of time and grievance settlement must happen fast or not at all. Therefore, they often prefer direct action (wildcats and picket lines) if there is a dispute about the application of the contract.

clause in the contract; prior to the signing of the present agreement the union had to collect the dues itself as best it could. The union wants a union shop (union membership a condition of continued employment).

Article 6: Management Functions

This is a very contentious issue. The present Duro clause gives broad powers to management and leaves the initiative in management's hands. The role of the union is that of the policeman—to see that the employee rights which have been bargained in the past are not jeopardized by on-going management decisions. (In Canada, generally, management is free to act unless such action is constrained by the intent of express clauses in the contract.)[4]

The practice of the company's contracting out considerable work has antagonized the union and the employees for some time; indeed, the company's way of handling layoffs and transfers resulting from contracting out was a contributing factor to the union's getting into Duro in the first place. A few months ago the union took a grievance to arbitration objecting to the layoff of an employee when the company contracted out work he had been doing, but the union lost on the grounds that there was no explicit or reasonably implied limitation on this practice in the contract. The union wants express language in the contract to limit management's freedom of action in this regard. Management, of course, would prefer to leave the clause in its present form.

Articles 7, 8, and 9: Hours of Work, Overtime, Allowances, Holiday and Vacation Pay

The union wants improvements in as many as possible of these benefits, including the following.

Article 7.02 Union to be consulted and agree to any change in regular hours of work.

---

4. Even if there is disagreement, management acts and gives instructions; it is up to the employee to carry out these instructions (except if there is a risk to his personal safety). Any arguments are settled through the grievance procedure (including arbitration, if necessary). If higher levels of management or arbitration reverse decisions made at low levels, then the company restores the situation to the way that it was before the grievance was filed.

Article 7.03 Time-and-a-half for hours in excess of eight per day. Overtime shall be equitably distributed, as far as is practical, among those normally performing the work.

The issue of compulsory overtime has been a troublesome one at Duro for the past two years. In the absence of specific language in the agreement to the contrary, management, under Article 6: Management Functions, has the right to insist that employees work overtime whenever requested. The union claims that management makes little or no attempt to give employees reasonable notice of overtime; further, the union feels strongly that an employee should have the right to refuse overtime if he wants to. The union wants specific language in the new agreement to provide for: (a) adequate notice; and (b) the right to refuse overtime. The company wants no change; overtime is important at Duro, particularly now, since big orders tax their physical facilities to the limit at times.

Article 7.04 Longer than twenty minutes for a lunch break.

Article 7.05 More than three hours call-in pay.

Article 8.01 More than eight statutory holidays with pay.

Article 8.02 Drop the limitation on statutory holiday pay.

Article 8.03 Better than double-time holiday pay.

Article 9.01 Much improved vacation benefits.

Article 10: Seniority, Section 10.04

During the last negotiations the company dug in on the question of seniority versus competence and qualifications, feeling that emphasis on seniority jeopardized efficiency. They were especially concerned because the more cyclical business associated with their emphasis on large orders for institutions and big office buildings, the more layoffs and transfers resulted. This issue almost caused a strike but at the last minute the company agreed to the procedure in the present contract because the union was adamant. The union, in turn, reduced its demands for better vacations and union security and also went along, though with considerable misgivings, with the company's proposal for an incentive plan.

Note that the company gets *adequately* qualified people, not the best qualified. Note, too, that bumping can take place in the department affected and even outside that department if necessary. Taken together, these two clauses give a great deal of job security to Duro's long-service employees. In this respect the Duro contract is not unusual; contracts frequently give more weight to seniority in layoffs than in promotions because so much can be at stake for the employee.

Since the introduction of this procedure at Duro, there have been relatively few grievances regarding layoffs. In some cases management has felt that a better qualified man would have been retained but for this procedure and would prefer to revert to less emphasis on seniority, as is the case with promotions and transfers.[5]

Management would like this clause changed so as to give greater weight to qualifications and competence and less to seniority. They now realize that the expression "adequately qualified" has worked to their disadvantage. They would prefer a phrase such as "ready

---

5. The effect of seniority on efficiency is a contentious question. Some argue that productivity improves with the higher morale resulting from freedom from fear of unfair and arbitrary layoffs. Others argue, however, that efficiency suffers if the threat of discharge is reduced. The relative weight which you give to these opposing arguments depends partly on your management philosophy. If employees dislike work and can only be motivated by carrots and sticks, you probably see seniority as a serious threat. On the other hand, if you believe that most employees like to do a good job and that high productivity can result from spontaneous cooperation, then seniority is less of a threat. [For more on this topic, see Douglas McGregor, *The Human Side of Enterprise* (New York: McGraw-Hill, 1960)] Most managements want the right to emphasize ability and merit when promoting employees. They argue that seniority kills initiative because they cannot reward high performance and penalize low performance. Whether or not this argument "floats", however, depends largely on the actual (not the imagined) competence of management. If, in fact, employees who are head and shoulders over the others, *and are seen to be so,* are promoted, then more seniority may well lower productivity. But if they reward low performers, whether because they are playing favourites or because they have incomplete information, then greater emphasis on seniority could well be less damaging to morale.

Background Notes on Present Agreement

ability"; too many times the union has pressed the view that relatively poorly qualified but very senior employees, given reasonable time, could learn to do the job. They say that long-service employees deserve this protection.

Article 10: Section 8

Now that the union has persuaded the company (during the last negotiations) to agree to more weight placed on seniority in layoffs, they want the same thing in promotions and transfers. The company, of course, is quite happy with the present language because this allows them to reward demonstrated competence. During the past eighteen months, the union took to arbitration two cases where the company promoted junior men over senior men who had bid for the jobs. The union lost both cases.

Article 11: Incentives

Incentives were first introduced at Duro as a result of the last set of negotiations. Though there have been many minor disputes between the union and management regarding incentives for individual jobs, both parties are fairly well satisfied on the whole, with the operation of the plan and with the wording in article 11. To date, the union has not attempted to take an incentive grievance to arbitration.

The company feels that this procedure is a big factor in getting higher efficiency. They took the initiative in proposing the plan only after very careful study and consultation with experts in other companies subscribing to incentive plans—some of which were good, some bad.

Article 12: Grievance Procedure

Both the company and the union prefer that the language in this article remain unchanged.

Article 13: Wages

Basic wage rates increased only slightly as a result of the last negotiations. The union fought for a bigger increase on the grounds that:

(a) many employees would not benefit from the incentive pay program; and
(b) that incentive pay cost the company nothing, since additional production resulted or there was no extra pay.

The company took the position that employees wanting higher pay could bid on incentive jobs as they became vacant. The company believed that the employees would accept a small basic wage increase because of the introduction of the incentive program. They proved to be right; though some employees objected bitterly, and the union bargaining committee was resentful, the majority of employees voted for it at the ratification meeting.

The union committee is determined that this time there will be a substantial increase in base rates.

Article 14: Hospitalization, Medical, and Pension

The employees are very dissatisfied with these. In many other unionized companies having employees doing similar jobs, these benefits are much better and the company contribution to the cost about double that of Duro's. Indeed, the union made much of this point during the organizing campaign in the fall of 1969; it said that Duro employees were second-class citizens.

The union was not able to bargain any substantial improvements in these benefits into the first two contracts negotiated and it is determined that worthwhile gains must be made this time.

Background Notes on Present Agreement

5
The Present Agreement

This Agreement between Duro Metal Products, Limited, Calgary, Alberta (hereinafter called the "Company") and Local 9999, United Fabricators of America (hereinafter called the "Union") is effective 31 March 1971.

ARTICLE 1: GENERAL PURPOSE OF THE AGREEMENT

The purpose of the parties hereto is to set forth herein the Agreement covering wages, hours of work, and other terms and conditions of employment to be observed, and to provide a procedure for the prompt and equitable adjustment of alleged grievances to the end that there shall be no interruption or impeding of work, nor work stoppages, nor strikes, nor other interference with production during the life of this Agreement.

ARTICLE 2: RECOGNITION

2.01 The Company recognizes the Union as the sole and exclusive bargaining agent for all employees in the Calgary plant save and except supervisors, security guards, office workers, and time-study workers.

2.02 The Company will deduct monthly Union dues from the pay cheque of each employee who is a union member who gives signed authorization to the Company for such deduction so long as: (a) this Contract is in effect; or (b) so long as the authority is not revoked by the employee in writing, whichever of (a) or (b) occurs first. Dues collected will be promptly transmitted to the secretary-treasurer of the Union.

2.03 Supervisors shall not normally perform work done by an employee in the bargaining unit except in the following cases:

(a) experimental work;
(b) demonstration work to train employees;
(c) emergency conditions;
(d) work which is negligible in amount; and
(e) work normally done by a supervisor even though similar duties are found in jobs in the bargaining unit.[1]

2.04 No employee shall participate in Union activity on the Company's premises during his working hours or on Company time (save as expressly authorized by this Agreement, or by written permission of the Company). No meeting for Union purposes shall be held on Company premises except with the written permission of the Company. The Union will not distribute handbills, Union publications, or the like on Company premises except as permitted by the Company.

ARTICLE 3: REPRESENTATION

3.01 The Union may select departmental stewards whose duties shall be limited to adjustments of disputes in the department for which they were appointed, while such disputes are being processed through steps 1 and 2 of the grievance procedure. Employees so selected must have one year's service or more with the Company. Stewards must obtain the permission of the foreman before leaving the work place. This permission will not be unreasonably withheld.

---

1. The bargaining unit consists of all employees in jobs represented by the union (whether union members or not). The types of jobs to be included are agreed on by the company and the union or, in the event of disagreement, are decided at a hearing before the Labour Relations Board.

3.02 The Union may select a Grievance Committee of four members, whose duties are limited to adjustment of disputes at step 3. Committee members will be paid at their average hourly rate for the preceding pay period for attendance at regular and emergency meetings, if any, but not exceeding a total of ten hours each in any calendar month. Any required time off the job beyond ten hours will be paid for by the Union.

3.03 It is agreed that the Union may appoint a Negotiating Committee of four employees. In addition, the Union may also be represented in negotiating meetings by the international representative of the Union. Employees who are members of the Negotiating Committee will suffer no loss of pay for time spent during their regular working hours at negotiating meetings with Company officials, provided such time, in the opinion of the Company, is not excessive.

ARTICLE 4: NO DISCRIMINATION

The Company and the Union agree that neither will discriminate against any employee because of race, colour, sex, or national origin.

ARTICLE 5: STRIKES AND LOCKOUTS

5.01 There shall be no lockout by the Company and no interruption, work stoppage, strike, or any other interference with production by any employee or employees during the term of this Agreement.

5.02 Any employee who participates in any interruption, work stoppage, strike, sit-down, slowdown, or any other interference with production may be disciplined or discharged by the Company.

ARTICLE 6: MANAGEMENT FUNCTIONS

The Union recognizes that the rights of the Company include, but are not limited to: directing the working forces, which involves the right to discipline, discharge for cause, hire, transfer, and promote; making reasonable rules; and determining its field of operations, hours of work, materials purchased, products manufactured, methods of manufacture, type and location of machines used, schedules, and sequence of operations. The Company agrees that the exercise of these functions will not be inconsistent with express provisions of this Agreement.

ARTICLES 7, 8, 9: HOURS OF WORK, OVERTIME, ALLOWANCES, HOLIDAY AND VACATION PAY

Note: Agreements usually use detailed language for these articles. These are quite explicit in practice, but for the sake of brevity those appearing in the Duro contract are simply listed below.

7.01 States starting and finishing time of regular work shifts and work week.

7.02 Provides forty-eight hours' notice to Union if above changed by Company.

7.03 Time-and-a-half for hours worked over forty in a week.

7.04 Twenty-minute paid lunch break during the shift.

7.05 Three-hour minimum pay if employee called in for emergency.

7.06 Pay made up if employee called for jury duty.

8.01 Eight holidays—New Year's Day, Good Friday, Victoria Day, Dominion Day, Civic Holiday, Labour Day, Thanksgiving, Christmas—with pay at the base rate for the job.

8.02 Requirement to work day before and day after a holiday to be eligible for holiday pay.

8.03 Double time for holiday work.

8.04 Extra day's vacation if holiday falls in vacation period.

The Agreement

9.01 Sets out vacation period based on length of service:

Service	Paid Vacation (at base rate)
Less than one year	Half day per month of service (maximum vacation, one week)
One year or more	One week
Three years or more	Two weeks
Fifteen years or more	Three weeks

9.02 Vacation pay for employees terminating before taking vacation (based on pro-rated service during the year and at a rate of 2 per cent for less than three years, 4 per cent if over three, and 6 per cent if over fifteen).

9.03 Permits choice of vacation period determined by seniority insofar as circumstances permit.

ARTICLE 10: SENIORITY

10.01 Seniority means the employee's continuous length of service with the Company.

10.02 Seniority and employment shall be terminated when an employee resigns, is discharged, is laid off for lack of work for a period exceeding one year, is absent for more than three days without notifying the Company, or fails to report within six working days after a recall from layoff.

10.03 An employee shall be considered a probationary employee until employed for three months. No grievance may be presented in connection with the layoff or discharge of a probationary employee.

10.04 In the event of a reduction in the work force of a job classification (see page 53) the junior employee in the classification shall be removed from that classification, provided the senior employees retained are adequately qualified to perform the remaining work

The Agreement

required. Such a displaced employee will be eligible to displace a more junior employee in the same department if he is adequately qualified to do the job required. If this is not possible he may replace a more junior employee in the plant if he is qualified. To keep displacement (bumping) to a minimum under the above procedure, the Company will start at the bottom of the appropriate seniority list and work upwards. If no job is available under the above procedure, the employee will be laid off.

10.05 Displacement rights will not apply in the event of a temporary layoff (fourteen days or less).

10.06 When an employee has been laid off he shall be entitled to recall in inverse order of the layoff procedure.

10.07 Members of the Union Negotiation and Grievance committees and Union officers will not be laid off, notwithstanding their seniority, so long as there is work available that they are qualified to do.

10.08 When a vacancy occurs it will be posted on the plant bulletin boards for forty-eight hours. During this period interested employees may make application (bid). The Company will consider these applications carefully and base its decision on qualifications, ability, and seniority, in that order.

10.09 In the event that an employee is promoted from the bargaining unit to a supervisory position, he shall retain the seniority he had accumulated prior to his promotion but will not accumulate seniority after his promotion.

ARTICLE 11: INCENTIVES

11.01 The Company may, at its discretion, establish incentive standards for any plant operation where incentives can be applied to increase productivity and to provide opportunity for employees to increase earnings above the standard hourly-wage rate scale.

11.02 The base rate for incentive will be the standard hourly rate for the job classification.

11.03 Incentive standards shall be established so that an average employee experienced on the job and working at a normal pace shall earn the base rate plus proportionately more for production above standard. Earnings shall increase in direct proportion to increased output; that is, a 1 per cent increase in output over standard shall result in a 1 per cent increase in earnings over the base rate of the job classification.

11.04 Incentive standards shall provide an opportunity for an employee, experienced on the job, working in accordance with prescribed procedures and without fatigue to a degree injurious to health, to increase his earnings by an amount up to 25 per cent above base rate.

11.05 An incentive standard, when established, shall remain unchanged except when a change or accumulation of changes affects the standard to a substantial extent, such change or changes having resulted from: a change in methods, materials, tools, equipment, inspection standards, or an error in computation in setting the standard.

11.06 Whenever the Company changes an existing standard or develops a new one, the standard shall be installed on a trial basis for seven days, at the end of which time either it will be accepted as standard or the employee may register a grievance claiming that it is not in accordance with this article.

11.07 An employee on an incentive job will not earn less than his base rate multiplied by the total clock hours worked on incentive.

11.08 An employee on incentive shall not earn incentive for work which was rejected because it did not meet inspection standards unless the cause of such rejection is beyond the employee's control.

11.09 A designated Union representative shall, on request, be supplied with

The Agreement

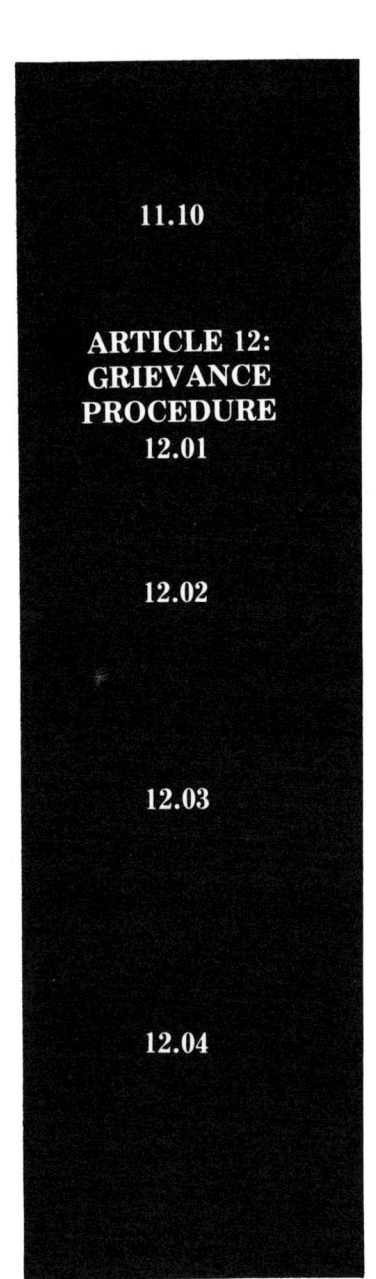

the material and data used by the Company in the establishment of a standard. This Union representative shall be permitted to view an operation which is in dispute.

11.10 An allegation of inaccuracy based solely on a claim of inadequate earnings shall not be grounds for disputing a standard except when the employee consistently earns less on a specific job than his established earning pattern on other incentive jobs.

ARTICLE 12: GRIEVANCE PROCEDURE

12.01 *Step 1* Any employee who believes he has a grievance may discuss and attempt to settle it with his foreman with or without the presence of the steward.

12.02 *Step 2* If a satisfactory answer is not received at step 1, the steward and/or the Grievance Committee may, within three working days of receipt of such answer, present the grievance with the facts in writing on the proper grievance forms to the production manager, who will give his decision within two working days of receipt of the grievance.

12.03 *Step 3* If a satisfactory answer is not received at step 2, the Grievance Committee may within three working days notify the Company of its desire to meet with the Company to discuss the grievance. A representative of the International Union may be present at such meeting. The Company will answer the grievance within four working days of the date the meeting at step 3 is held.

12.04 *Step 4* If the grievance is not settled at step 3, the Union may, within thirty calendar days of receipt of the answer at step 3, notify the Company of its intention to submit the grievance to arbitration. The decision of the majority of a three-man Arbitration Board or single arbitrator shall be final and binding on both the Company and the Union. The Company and the Union shall pay one-half the remuneration of the single arbitrator or the chairman of the Arbitration Board and shall each bear its own expenses.

12.05 Any grievance not processed from one step to the next by the grieving party within the time limits specified above, unless such time limits are mutually extended in writing, shall be considered settled on the basis of the last answer given.

12.06 The Grievance and Arbitration Procedure may be invoked by the Company at step 3 of the Grievance Procedure. For such purposes this section will be read with necessary changes.

12.07 An employee who believes he has a grievance shall process it in the manner provided within seven working days of the occurrence which gave rise to the complaint or within seven days of the date the employee or the Union should reasonably become aware of the occurrence which gave rise to the complaint.

ARTICLE 13: WAGES

13.01 Effective 17 March 1971, the wage rates listed under Schedule A below will become effective and remain in effect until 31 December 1971. Effective 1 January 1972 and until 31 December 1972, wage rates under Schedule B will apply.

Classification Number	Classification	Schedule A	Schedule B
1	Janitor, Degreaser, General Labourer	$2.80	$2.90
2	Packer, Crate Maker, Drill Press Operator, Spot Welder, Press Operator	2.90	3.00
3	Shipper, Oiler, Handtruck (battery)	3.05	3.15
4	Tow Motor Operator, Paint Mixer	3.19	3.30
5	Crane Operator, Expediter	3.33	3.45
6	Truck Driver, Inspector, Arc-Acetylene Welding	3.48	3.60

The Agreement

Classification Number	Classification	Schedule A	B
7	Machinist (2), Carpenter, Shop Lead Hand	$3.63	$3.75
8	Machinist (1), Electrician A	3.83	3.95
9	Toolmaker (1), Maintenance Lead Hand	4.06	4.20

13.02 It is understood that the Company retains the sole right to alter, add, or discontinue jobs. If a new job is added the Company will establish a new classification commensurate with the present rate structure.

13.03 A shift premium of ten cents per hour will be paid for all hours worked on an employee's regularly scheduled afternoon shift (3:00 P.M. to 11:00 P.M.).

13.04 The Company may require an employee to take a temporary job other than the job on which he is regularly employed, for any period not exceeding two weeks, provided that he receive the rate of pay for his regular job or the rate of pay for the new job, whichever is higher.

ARTICLE 14: HOSPITALIZATION, MEDICAL, AND PENSION

Note: In addition to the clauses and benefits in the preceding articles, the agreement has other fringe benefits for Duro employees. These are summarized, in simplified form, below.

The Company and the employees share equally the cost of:

(a) $5,000 life insurance for each employee;
(b) hospitalization benefits for employee and dependents;
(c) surgical benefits for employee and dependents;
(d) weekly indemnity payments if employee unable to work because of sickness or accident, in an amount approximately equal to one-half of basic earnings, for a maximum period of thirty weeks, payable beginning the first day of accident and the fourth day of sickness.

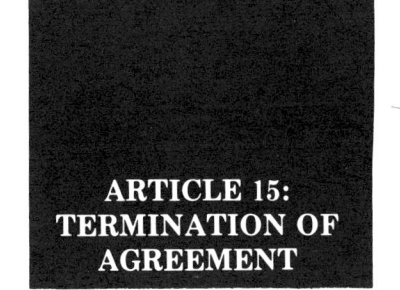

The Company also administers a plan which pays for doctor visits. This is fully paid for by the employees.

The Company contributes, as required by law, to the Canada Pension Plan. No other pension is provided.

ARTICLE 15: TERMINATION OF AGREEMENT

This agreement shall be in effect until 31 December 1972, and continue thereafter until either party gives the other notice in writing of a desire to terminate or amend it.

The Agreement

6
Costing the Package

Your team score will depend *primarily* on how far the added cost over three years deviates from the optimal positions shown in figures 1.1 and 1.2 (pages 6 and 7).

In addition to points lost due to the above deviation, both union and management teams lose points if there is a strike. This penalty affects your score in the same way as would an additional wage deviation of 4¢ per hour (12¢ in third year).

Following are assumptions you may use in calculating the added cost of certain benefits if you change them. If you deviate from these assumptions it must be by agreement of both management and the union.

7.03: Overtime Rate

Assume the average amount of overtime worked by each employee is six hours per month.

7.05: Call-in Pay

Assume this occurs rarely; say, twice per year per employee. Assume, too, that extra paid hours (if any) return no extra work to the company.

8.02: Holiday Qualification

If you remove this qualification on holiday pay, assume a 3 per cent loss in productivity during the period beginning the day before through to and including the day after (a total of 6 per cent of a day's pay, in all, per holiday).

Costing the Package

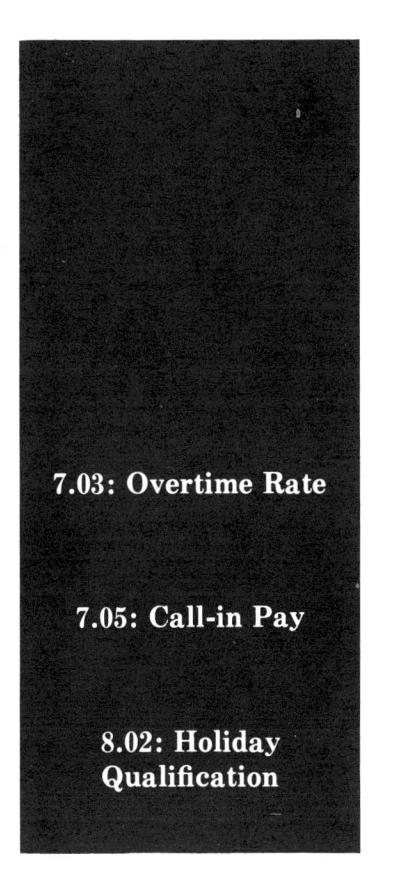

8.03: Overtime on Holidays

9.01: Vacation Pay

Assume this occurs rarely; that is, once per employee in a two-year period.

The employees have the following service with the company:

Years of Service	Number of Employees
Less than one	20
One but less than three	40
Three but less than five	30
Five but less than ten	10
Ten or over	4

13.01

If you bargain, say, a 1 per cent increase, basic wages by the third year will increase the average basic wage by 3 per cent, from 3.27 to 3.37 (3.3681 rounded off).

Note: In addition to the above, incentive wages will increase also. As stated earlier, 60 per cent of the employees in the bargaining unit are on incentive and they average 20 per cent above the base rates.

Article 14

Report, for each separate benefit, any added cost of changes, in terms of cents per hour equivalent. If, for instance, you bargain more life insurance which is to cost the company $20 per year per employee, this would be reported as 1¢ per hour (using 2,000 hours per year for ease in calculations).

If time permits and your instructor so indicates, you may, of course, bargain exact benefits indicating the details and cost. Information on costs can be obtained from personnel departments of local firms and from insurance companies.

Costing the Package

7

Three Selected Issues

PROMOTION AND TRANSFER CLAUSE

Alternative Basic Clauses

Optional Additional Clauses

Bargain one of the basic clauses (a) through (d), adding, if you wish, optional clause(s) 1(a) *or* 1(b), and/or clause 2. Each team member is to report (on the Negotiation Report, appendix V) the clause(s) agreed upon, or, if there is no contract agreed upon and there is a strike, the final position of his team on this issue.

When a vacancy occurs it will be posted on the plant bulletin boards for forty-eight hours. During this period interested employees may make application (bid). The Company will consider these applications carefully and base its decision on:

(a) (same as clause in the present agreement—10.18) qualifications, ability, and seniority, in that order;
(b) qualifications and ability, provided that when these are substantially equal, preference will be given to the applicant with the most seniority;
(c) seniority, provided that the applicant has the ready ability and qualifications to do the job in an efficient manner;
(d) seniority, provided that the applicant can demonstrate the aptitude for learning the job.

1. After a reasonable trial period, if the successful bidder (if any) is unable to do the job efficiently, he will:

(a) return to his former job and will not be permitted to bid on another job for a period of six months; or

Three Selected Issues

(b) have the choice of open jobs (jobs for which there are no qualified bidders) which, in the opinion of the Company, he can do efficiently, or, if there are no such jobs, he may displace the least senior employee in the plant whose job he can do satisfactorily. In either case, he cannot bid on another job for a period of six months.

2. For the purposes of transfers and promotions, only biddable vacancies will be posted. A biddable vacancy is defined by and limited to the following:

(a) the job vacated by an employee who quits, retires, or is discharged;
(b) the job vacated by the employee who takes job (a);
(c) the job vacated by the employee who takes job (b).

In the above cases, if a qualified applicant is not available from within and the company hires a new employee, this breaks the sequence and no further vacancies are posted. In no case will there be more than three jobs posted as a result of one quit, retirement, or discharge.

UNION SECURITY CLAUSE

Bargain one of the following clauses related to union security. Each player is to report (on the Negotiation Report, page 89) the clause agreed upon, or, if there is no contract agreed upon and a strike results, the final position of his team on this issue.

Clause 1

(same as clause in the present Agreement—2.02)
The Company will deduct monthly union dues from the pay cheque of each employee who is a Union member and who gives signed authorization to the Company for such deduction so long as: (a) this contract is in effect; or (b) so long as the authorization is not revoked by the employee in writing, whichever of (a) or (b) occurs first. Dues collected will be promptly transmitted to the secretary-treasurer of the Union.

Clause 2

The Company agrees that all employees who are or who become Union members shall remain dues-paying members for the duration of this Agreement as a condition of continuing employment. Each month the Company will deduct Union dues from the pay cheque of Union members who give signed authorization (on the form provided) for such deduction, and will promptly transmit such collected dues to the secretary-treasurer of the Union.

Clause 3

The Company agrees that all employees who are or who become members of the Union shall remain dues-paying members for the duration of this Agreement as a condition of continuing employment. Further, each month, the Company will deduct from the pay cheque of each employee in the bargaining unit, whether a member of the Union or not, an amount equivalent to monthly Union dues as specified in the Union Constitution, and will promptly transmit such deductions to the secretary-treasurer of the Union. Each employee, as a condition of continuing employment, will sign (on the form provided) to authorize such monthly deduction.

Clause 4

Employees in the bargaining unit shall, as a condition of their continuing employment, become members of the Union during the first thirty days of their employment with the Company or on the effective date of this Agreement, whichever is later, and shall continue membership in the Union during the period of this Agreement. The Company agrees to deduct monthly Union dues (as specified in the Union Constitution) from the pay cheque of all employees, each of whom will sign (on the form provided) to authorize such deduction. Funds deducted will be promptly transmitted by the Company to the secretary-treasurer of the Union.

Clause 5

The Company agrees that, for jobs in the bargaining unit, it will hire only employees who are members of the Union. The Company will deduct monthly Union dues and all employees will sign authorization forms. Such deductions will be promptly transmitted by the Company to the secretary-treasurer of the Union.

CONTRACTING OUT

Bargain one of the clauses below. Each player is to report the clauses agreed upon, or, if no final contract is agreed upon, report the clauses

Three Selected Issues

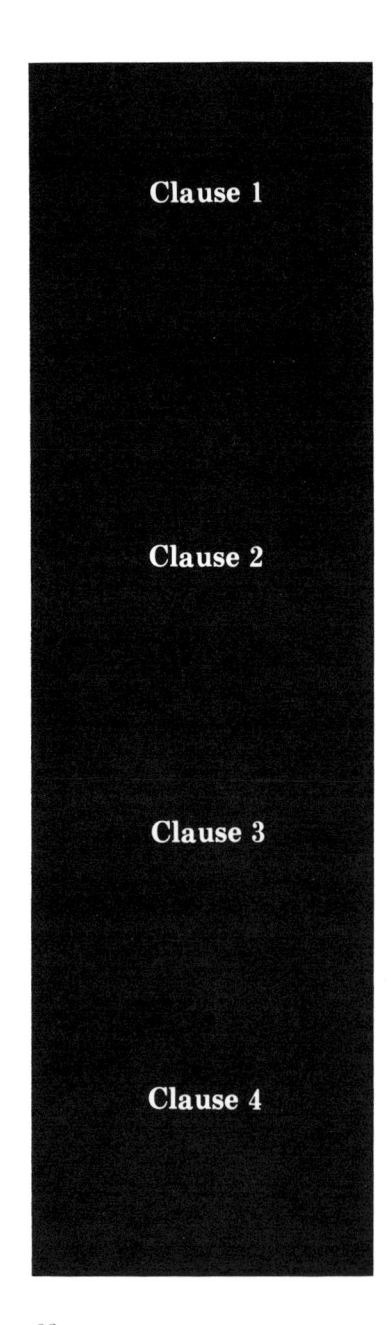

last offered by his team (on the Negotiation Report, page 89). Bargain one of 1, 2, 3, 4, 5 *plus* one of (a) or (b).

Clause 1

(same as clause in present contract—article 6)
The Union recognizes that the rights of the Company include, but are not limited to: directing the working forces, which involves the right to discipline, discharge for cause, hire, transfer, and promote; making reasonable rules; and determining its field of operations, hours of work, materials purchased, products manufactured, methods of manufacture, type and location of machine used, schedules, and sequence of operations. The Company agrees that the exercise of these functions will not be inconsistent with the express provisions of this Agreement.

Clause 2

(same as 1 but adding the following)
Before the letting of outside contracts for work normally performed by employees in the bargaining unit, or before purchasing, in a prefinished state, parts previously manufactured in the plant, the Company will, as far in advance as is reasonably possible, inform the Union as to the reasons for such action and the probable effect on the type and number of jobs in the bargaining unit.

Clause 3

(same as 1 but adding the following)
Except in the cases of an emergency likely to jeopardize delivery commitments made to customers, the Company will not contract out work normally performed by employees in the bargaining unit if such action results immediately in a layoff, the failure to recall laid-off employees, or the offering of less than a normal work week to bargaining unit employees.

Clause 4

(same as 3 but adding the following)
If the employee elects not to accept such transfer, he has the option of quitting and receiving special displacement benefits as follows: 100 hours' pay at his regular hourly rate for each full year of service with the Company (credit for partial year's service to be on a pro-rata basis).

Clause 5

Additional Clauses

(same as 1 but adding the following)

Before letting outside contracts for work normally performed by employees in the bargaining unit, or before purchasing, in a prefinished state, parts previously manufactured in the plant, the Company will consult with and obtain the agreement of the Union as to the resulting effect of the action on: (a) the types and numbers of jobs in the bargaining unit; and (b) the rates of pay and the personnel to staff the remaining jobs.

(a) In the event that an employee is transferred by the Company because his job has disappeared, the employee will receive the customary rate of pay of the new job to which he has been assigned.

(b) In the event that an employee is transferred by the Company because his job has disappeared, the employee will receive either the rate of pay of the new job to which he has been assigned or the rate of pay of his former job, whichever is higher. In the event, however, that his rate exceeds the job rate, he will not receive general wage increases until such time as the job rate equals or exceeds his rate.

Appendices

I Contract Ratification by Union Members

Very often the constitution of the union provides that the final offer of the company be voted on by the general union membership. Sometimes the union members refuse to accept the best deal which their committee says they can get, and they send the committee back to management to ask for more.

This can be very embarrassing if the committee has shaken hands with management on a package, honestly believing that the members would accept it. And it can lead to serious trouble since management may well have held nothing back to deal with this eventuality. This is a situation where management sometimes prefers strong union leadership to weak. With weak leadership the union is often divided internally, with aggressive factions trying to oust the present leaders by getting into office at the next election. Their outspoken criticism may prevent acceptance of *any* reasonable management offer.[1]

Other times, the committee, even though not satisfied with the last offer of the Company, may agree to present it to the members for a vote. In such cases the method of presentation can make quite a

---

1. This may well be an important factor in the current troubles with the postal union in Canada. When, a few years ago, the leadership changed, the new leaders were not able to get the broad base of acceptance necessary to keep the various factions in line. The Teamsters provide another example. Hoffa, whatever else is said of him, was in control of the union and when he said, "It's a deal," it did not become unstuck.

difference. In one case known to the writer, the chief negotiator for the Union stood in front of the general membership meeting and shouted, "Let's show them what we think of their offer: here's what I think of it," whereupon he ripped the paper into shreds, dropping them in a pile at his feet, then jumped up and down on them.

II Divergent Problem Solving and Collective Bargaining

Different approaches are needed to solve different types of business problems. For instance, the problem-solving approach used by engineers to design a bridge will not work in collective bargaining.

It is useful to separate problem situations into two types, convergent and divergent. Convergent applies when you have enough information at the outset to converge on the optimal solution. This is the traditional scientific method. Typically there are only a few variables and these are known; the variables do not interact, and cause-and-effect relations hold. Alternate solutions to problems can be analyzed and evaluated in advance and the one best sequence of actions selected. This approach is useful for many business problems: production scheduling, inventory control, optimal plant location, batch mixing, and structural design, to name a few. Many people seem to be temperamentally suited to this type of problem solving and work best in situations requiring this approach. The convergent approach permits effective planning and provides a master blueprint, a program for all to follow. Not all situations, however, lend themselves to this. Often, dangerous oversimplification results because of the concentration on what can be counted or measured; not all important factors are quantifiable. Further, since precise quantitative methods are usually involved, there may be an illusion of validity which does not necessarily hold true. Some industrial problems are simply too complex to plan a complete sequence of actions far in advance, so that once started, thousands of small parts will click smoothly into place.

Complex situations often require a divergent approach. Here, there are many known and unknown variables, all interacting with one another. Relations are best stated in terms of probabilities instead of cause-and-effect. There is a great deal of uncertainty, both because not all the desirable information is available, and because the situation is dynamic. The divergent approach involves carefully collecting a feasible amount of information, analyzing it, making a commitment, and then acting and observing the outcome. The results (output) of this act then become part of the input for a new decision, and the cycle is repeated as often as desired. Often there isn't time (or tools either, for that matter) to develop perfect and certain solutions. You satisfice rather than maximize.[1] In these situations, if management fiddled around waiting for perfect solutions (perhaps so that they couldn't later be accused of making a mistake), they would wait too long. By the time they had acted, the underlying factors would have changed to the point where they would have to start over; they would never act.

This simulation involves divergent problem solving. You will not have all of the information which you would like about the company. You will have to make inferences from what your opponents say. And at times you must make good decisions quickly to grasp opportunities and not cave in under pressure. You must continually reassess your strategy in the light of developing situations and new information.

1. Researchers in organizational behaviour, such as R. Cyert and J. March, point to the tendency for decision makers who must make many decisions in a limited amount of time to "satisfice"—to select an option that is good enough—in order to get on to the next point of discussion.

III Industrial Relations in Canada Today—An Overview

Collective bargaining is joint decision making by management and unions on matters of mutual interest, particularly wages, fringe benefits, job security, job opportunities, employee grievances, and working conditions. Employees select the union which the majority wants, then elect fellow employees as union officials to represent them in contract negotiation and contract administration.

It is important to note that collective bargaining is public policy in Canada and since the 1940s has been encouraged by legislation. Employers, their personal philosophies notwithstanding, must by law: (a) recognize and bargain with the union of the employees' choice; and (b) bargain in good faith. This legislation reflects the view of the majority of Canadians that collective bargaining is the best way to encourage a good standard of living for Canadian employees while at the same time providing machinery for reconciling the conflicting interests of employees and employers.

In Canada, except for a few special industries such as shipping and railways, legislation on labour relations comes under provincial jurisdiction. Other countries handle this in different ways. In the United States it is federal; in the United Kingdom the situation is so chaotic that one wonders if anyone has jurisdiction. There are some differences in the methods of legislation of the various provinces, but in most respects they are similar. In all cases there is an element of compulsion and considerable regulation of the relationship between a company and a union. Generally speaking, the following hold:

(a) compulsory and binding arbitration of disputes arising during the term of a contract;

(b) compulsory but not binding (except in British Columbia) conciliation of disputes involving the terms of a new contract; here government conciliators attempt to bring the parties together and prevent a strike or lockout;

(c) prohibition of strikes and lockouts during the term of an agreement, or during the period that government conciliators are working with the parties negotiating a contract;

(d) provincial labour boards to enforce legislation and hold hearings if either management or the union alleges that the other has violated the law. These boards also "certify" a particular union to represent the employees in a firm, once the union establishes that a majority of the employees want it. When a union is certified for a period of time, no other union can represent the employees.

The labour legislation of the provinces attempts to maximize productivity by minimizing work stoppages and slowdowns. We are one of the most industrialized nations in the world and our industrial complex is highly integrated. In order to plan, companies must have dependable sources of supply for raw materials, machinery, building material, labour, transportation and other services. A work stoppage in one plant can interfere with deliveries to another plant, thus reducing efficiency; and in Canada, efficiency is important. In the first place, over a quarter of those in the Canadian work force are dependent on export-based industries for their jobs. Export markets are competitive and our prices must be right. Further, a high standard of living (which comes from high productivity) may well be a prerequisite of nationhood, for without it, too many of our highly trained people will be drained off to the United States, attracted by the higher incomes there. We cannot build a Berlin Wall to stop this, as the East Germans had to do—a 3,000-mile-long birchbark barricade would look pretty silly!

Seen in perspective, collective bargaining has worked well in Canada, notwithstanding some strident claims to the contrary.[1] Critics point to the time lost due to strikes, wring their hands at reports of picket-line violence, and bewail spiralling inflation, which they attribute largely to unreasonable union demands. There is some truth in all of this, of course, but no system peopled by human beings will ever be perfect. The critics have yet to come up with an alternative system that is likely to be any better.

Statistics have shown that millions of man-hours are lost each year due to strikes, and this is obviously bad. But such absolute figures are dangerous, for they only tell part of the story. In a typical year, the time lost due to strikes represents less than half of 1 per cent of the total hours worked, a figure which, perhaps, should be increased to 1 per cent if indirect effects (in plants not on strike) are included. It is easy to form exaggerated impressions of the damage done by strikes. Strikes make good news copy; peaceful settlements do not. Newspapers compete for readers; understandably, then, they print what their readers want to hear. Strikes make the headlines; settlements, if they are mentioned at all, are often summarized in only a few lines in the back pages. But for every contract settlement involving strikes and violence, there are dozens settled less dramatically.

Unions, too, usually appear to be the aggressors; it is they who take the initiative (by striking), not the management (by locking out). This, of course, will always be the case in a rising economy; the status quo favours management. In a falling economy management would be taking the initiative and more of the blame.

The issues involved in collective bargaining (costs and working conditions) are of life-and-death importance to both employers and

---

1. Management spokesmen, naturally, would like to see the bargaining power of the unions reduced by curtailing the union's right to strike. These same spokesmen, however, are cool to any suggestion that their right to set product prices be curtailed in any way.

employees. There will always be many disputes, misunderstandings, and conflicts of interest. There is no reason to believe that there would be fewer stoppages if they were allowed to take place whenever the company or the union felt like it.[2] As it is now, the parties bargain a contract for, say, three years, and they are stuck with it. Neither side can alter the terms of the agreement unilaterally: management cannot reduce benefits; the union cannot improve them. Both sides learn the importance of arriving at clear and workable understandings during negotiations. Important issues are given attention by top-level management and union leadership, instead of being swept under the rug to smoulder.

Then there is the question of picket-line violence. Certainly it happens, and using violence on people who are trying to cross picket lines when they have the legal right to do so is clearly wrong. But violence can take many forms and violence begets violence. A nonunion employer, for instance, has the legal right to fire an employee without cause with notice, or with cause without notice. This means that an employer, in a moment of pique, has the legal right to fire a long-service employee, provided he gives, say, a week's pay or notice, just because he did not like his looks or his attitude; and the employee has no recourse whatsoever. If such an employee is middle-aged and has no special skills he may never get a decent job again as long as he lives. Many would call this a form of violence too. Violence, no doubt, takes place on picket lines outside the plant gates, but without effective picket lines, violence of a different sort may well be more prevalent inside the plant gates. The problem is how to eliminate one form of violence without increasing the other. The law-and-order advocates who keep asking each other where it is all going to end are silent on how to cope with situations where the law is outmoded or inadequate.

---

2. Nor can pervasive and genuine grievances be simply legislated out of existence by passing a law to forbid strikes. This is something like trying to nail jelly to the wall; instead of being manifested in the overt form of legal strikes, dissatisfaction can take more covert forms—wildcats, slowdowns, sabotage, reporting sick, and lack of cooperation generally. Sometimes, the reaction to a law is outright defiance: witness the experience of the U.S.A. in trying to stop drinking by prohibition.

Critics also blame collective bargaining and unions for inflation. They call this cost-push inflation and say that increased labour costs are passed along to the consumer in the form of higher prices. Unions do not agree that they are the main culprits. They point to the rising incomes of managers and professionals, to increasing profits, and to steadily increasing productivity. They ask how many of the critics, when granted their last raise, went to the boss with the suggestion that it be cut in half as a means of lowering costs and fighting against inflation. Apparently labour group members, like others, are reluctant to be the first to sacrifice or pass up opportunities to get ahead as long as they feel others are not doing so.

There is no doubt that employees who bargain as a group have more power than they would have as individuals, so hourly *rates* are indeed higher because of unions. It does not necessarily follow, however, that labour *costs* are higher, if this pressure forces management to continually become more and more efficient in the use of labour, as is sometimes the case.

From this brief review of the conventional charges levelled against collective bargaining, it seems that the case against collective bargaining leaves something to be desired; on the whole, collective bargaining serves this country well. We are, after all, one of the most productive nations in the world and we have one of the highest standards of living. Most countries would dearly love to have our problems!

This does not mean, though, that collective bargaining faces no serious challenges. There are soft spots which need attention. The private decisions made by companies and unions have public consequences, and public disenchantment could result in across-the-board legislation which would unduly weaken collective bargaining.

Three cases come to mind where many feel that the balance of power is weighted too much in favour of the union. These are:

(a) the skilled trades;
(b) unions in semimonopolistic industries; and
(c) unions in vital industries such as hospitals and utilities.

To the skilled trades must go the credit (or blame, depending on your point of view) for keeping the union movement alive in the last century and the first third of this one, thus setting the stage for the take-off of the industrial unions in the 1930s. Only they had the necessary bargaining power and the necessary, almost missionary, zeal. But it is this very bargaining power and the uses to which it is put that contributes to the bad press of the whole union movement. Skilled trades have control over the number of tradesmen trained; not surprisingly, therefore, they are not generally in oversupply, and they are not easily replaced. (In this respect they are much like the professional associations of doctors, lawyers, and dentists.) Further, the trades are fragmented; in construction, for instance, an employer sometimes must deal with several different unions, each with only a few key men on the job. When such unions take turns honouring each other's picket lines, massive construction projects are held up at great cost to the builder and to the public. They therefore have the power to inflict great damage at relatively little cost to themselves. As a result, some have bargained exceedingly high hourly rates. Further, brotherly love does not extend beyond a particular union and the various unions fight bitterly over who is entitled to do what. Wildcat strikes and "jurisdictional" disputes delay projects and reduce efficiency. For these reasons the skilled trades are sometimes called the "Achilles' Heel of the union movement".

In the case of semimonopolistic industries, unions also have a great deal of bargaining power based on the employer's ability to pay. In such industries there are opportunities for price collusion between employers, and high labour costs can be passed along to consumers in the form of higher prices more readily than in highly competitive industries. Such unionized blue-collar workers receive high wages; in fact the wages may well be reaching the point where there is a dangerous gap between them and less favoured groups in the population, especially nonunion workers, the unemployed, and retired people. Employees in strong unions are no longer underdogs in society; they are part of an industrial elite.

Thirdly, there is the problem of unions in vital industries. A relatively small number of Seaway employees, for instance, could just about hold up the country for ransom (even a substantial wage increase would be nothing compared to the cost to the country of a prolonged Seaway tie-up). It is true that unions in these industries often do not exercise their full power because of the fear of public retaliation through legislation; nevertheless, they are a source of worry. It may not be fair to deny them the right to strike, since this is the foundation of all unions' bargaining power; certainly they feel this would reduce them to the role of second-class citizens. On the other hand, when garbagemen and hospital employees do strike they alienate large sections of the population.

The three situations described above do create problems, but hamstringing the institution of collective bargaining would be somewhat like using a cannon to hunt rabbits. Particular, not general, solutions are needed for particular problems. Some, for instance, suggest that compulsory mediation should be substituted for the right to strike if the parties cannot agree on the terms of a new contract. But making strikes illegal does not prevent other forms of direct action; tying down the safety valve will not prevent a boiler explosion. And there is another problem. Though compulsory mediation may start by *complementing* collective bargaining, it might well end by *replacing* it. Like pregnancy, there may be no such thing as a little mediation. Knowing that mediation is always in the background, one party or the other might feel that mediation results would likely be better than what they were getting at the bargaining table, so that many disputes would go to mediation. Both sides would know that mediators look for the area of settlement somewhere between the last positions of the two sides. Mediation might well introduce a force that would widen, rather than close, the gap between the parties.

Facile solutions to very complex problems solve particular problems but create others. At the present time, any legislation that reduces the bargaining power of all unions would be a mistake. Many unions are too weak as it is.[3]

3. Many unions are weak financially and do not have the resources to organize certain industries, particularly industries made up of many small businesses that are often run by determined anti-union owners who fight unions step by step. In these situations union costs often exceed any potential revenues from dues. In this connection it should be noted that in a typical year, the total revenue of all unions in the United States and Canada is less than the profits made by General Motors in a few days.

Appendix III

There are several things that can be done. The skilled trades situation can be balanced by passing legislation to strengthen the hand of construction employer associations when they bargain with skilled-trade unions. This would tend to force the present fragmented unions into an integrated structure of one large and responsible union. High wages in semimonopolistic industries can be controlled by putting teeth into anticombines legislation at the federal level. With more competition in the product market, employers will resist (and very effectively) any excessive union demands. Finally, in vital industries new formulae will have to be found to assure the employees of fair wages where the ultimate weapon, the strike, is impractical. Agreements, for instance, can perhaps be negotiated which tie wages to, say, the average wage of five selected industries in the private sector.

IV The Office Furniture Industry in Canada

The hourly rated employees in the furniture industry are generally males. Most of the jobs are what might be called semiskilled. Training is received on the job. The upper limit on quality is determined by machines, though careless machine operation will produce low quality. Since the machines and other operations are not linked together in one long assembly line going throughout the plant, employees can pace their own work. Experienced operators' work skills become highly developed and with little or no added effort they can produce well above the level of beginners.

The sales of this industry fluctuate considerably, depending on the health of the economy as a whole. In 1970, for instance, sales increased only 2.3 per cent to $89.3 million, in sharp contrast to the preceding year when the increase was 20.8 per cent over 1968. Any uncertainty about economic prospects has a depressing influence on sales.

Statistics Canada: Catalogue 35-212, Office Furniture Manufacturers (1970), reports the following:

	Millions of Dollars
Wooden desks	$16.5
Wooden chairs	5.6
Wooden other	3.2
Metal desks	13.6
Metal chairs	11.9

	Millions of Dollars
Metal visible record equipment	$19.6
Value of shipments (1970) for total industry in Canada	$89.3

The product mix also varies from year to year. In 1970, sales of metal office desks declined 5.5 per cent, whereas sales of visible record equipment increased by 9 per cent.

The majority of firms in this industry are in Ontario and Quebec. Of some 110 firms listed in *Fraser's Canadian Trade Directory* only about a dozen are west of the Lakehead and a handful are in the Maritimes. Most firms are small, though there are a number with several hundred employees.

The outlook for the industry is good. White-collar workers constitute an increasing proportion of the work force, which is also growing overall.

The following pages contain statistics, some of which you may find useful to support your arguments when bargaining. You will notice that most industry statistics are for the metal fabricating industry[4]

TABLE IV.1 Average Weekly Earnings in Alberta (All Industries*)

Year	Average Earnings
1967 (July)	$101.53
1968 (December)	108.56
1969 (November)	121.51
1970 (August)	130.80
1971 (July)	140.70
1972 (February)	145.64

*Includes forestry, mining, durable and nondurable goods, construction, transportation, utilities, trade, finance and insurance, and services (both salaries and wage-earning employees).

4. This excludes machinery and transportation equipment.

(of which metal office furniture is a part). The primary source of data used is the *Canadian Statistical Review* (11-003), Statistics Canada.

TABLE IV.2   Average Weekly Earnings (Wages and Salaries), by Province, in Metal Fabricating Industry
(Nearest Dollar)

Year	Alberta	British Columbia	Ontario	Quebec
1967	111	118	112	111
1968	121	128	117	114
1969	133	146	125	119
1970	148	152	139	130
1971 (March)	149	157	142	129
1972 (February)	151	173	155	136

TABLE IV.3   Average Hourly Rate, by Province, in Metal Fabricating Industry
(Hourly Rated Wage-Earners)

Year	Alberta	British Columbia	Ontario	Quebec
1967 (July)	$2.57	$3.05	$2.55	$2.44
1968 (June)	2.92	3.37	2.76	2.61
1969 (December)	3.31	3.94	3.05	2.75
1970 (August)	3.51	4.05	3.37	3.02
1971 (March)	3.68	4.19	3.52	3.14
1972 (February)	3.89	4.55	3.77	3.29

TABLE IV.4   Annual Shipments By Metal Fabricating Manufacturers (Canada)

Year	Millions of Dollars
1967	$230
1968	235
1969	264
1970	268

Appendix IV

Year	Millions of Dollars
1971	$288
1972 (6 months)	307

TABLE IV.5 Office Furniture Industry Selling Price Index (Canada)
(1961 = 100)

Year	Total	Wood Office Desks	Metal Office Desks	Lockers and Shelving
1967	116	121	—	—
1968	116	125	—	116
1969	119	126	121	117
1970	127	132	130	127
1971	132	138	136	133
1972 (7 months)	137	—	—	—

TABLE IV.6 Consumer Price Indexes
(1961 = 100)

Year	Canada	Edmonton-Calgary
1967	115.4	111.3
1968	120.1	115.8
1969	125.5	119.7
1970	129.7	122.2
1971	133.4	123.4
1972 (July)	140.2	128.0

Note: This table can be used to calculate percentage increases in either column but *cannot* be used to compare Edmonton-Calgary with Canada directly (the base for both is 100, notwithstanding the difference which existed between Calgary-Edmonton and Canada in 1961, the base year).

V Team Strategy Report

(To be handed in to instructor after session 2.)

1. OPENING POSITION   (developed during session 2)

Note: List all changed items in your initial position, including, wherever possible, the cost in terms of cents per hour.

Wage increases are to be *annual*—if you demand or offer X cents per hour, this means that wages will increase by this amount each year of the three-year contract—but just report the X cents per hour.

Other changes will go into effect the first year and will continue but will not increase during the life of the contract.

Changed Item	Cost (Cents Per Hour)

2. INTERMEDIATE POSITION

Note: List below the approximate cost to the company of your intermediate and final positions:

(a) wages;
(b) *total* of the other readily costed items;
(c) a list of the items which are not readily costed, for example, changed seniority clauses.

Changed Item	Cost (Cents) Per Hour)
1. *Intermediate*	
2. *Final*	

Appendix V

Final Negotiation Report

Team No. _6_   Check one: Management _✓_
                           Union _____

| Agreement reached _____ | Agreement not reached (strike) _____ |

THREE SELECTED BARGAINING ISSUES

Note: Report the agreement reached, or, if there is a strike, your team's final position.

These three issues should be bargained as is—using only the alternatives supplied. Do not delete, add, or alter words.

1. Union Security Clause (2.02)

Circle one: A   B   (C)   D   E

NON-UNION EMPLOYEES MAY GIVE ___ DUES DEDUCTIONS TO CHARITY ___ ___ DO NOT WANT UNION TO HAVE IT

2. Contracting Out (article 6)

Circle one: Aa   Ba   Ca   Da   Ea

               Ab   Bb   (Cb)   Db   Eb

3. Promotion and Transfer (10.08)

Circle one: A   B   (C)   D   E

               Aa   Ba   Ca   Da   Ea

               Ab   Bb   Cb   Db   Eb

Optional Clause 2:   YES: _____   NO: _✓_

Appendix V   89

Note: Do not negotiate the following articles and sections of the contract:

1, 2.01, 2.03, 2.04, 3, 4, 5, 7.01, 7.06, 8.04, 9.03, 10.01, 10.02, 10.03, 10.05, 10.06, 10.07, 10.09, 11, 12, 13.02, 13.03, 13.04, 15.

Report your team's final position on the following; you may add, delete, or alter as you wish.

Article and Section	Check here if not changed	If different from present contract write the exact new wording below and indicate what is replaced. (Use margin of report when needed.)	Cost (Cents Per Hour)
7.02 Hours		THE COMPANY MUST GIVE 7 DAYS NOTICE IF CHANGE IS OF SEMI-PERMANENT OR PERMANENT NATURE - OTHERWISE 48 HR. NOTICE	—
7.03 Overtime	✓		—
7.04 Lunch Break		30. MIN LUNCH BREAK	12¢/hr/emp
7.05 Call-in Pay		MIN 4HRS CALL-IN PAY	.576¢/hr/emp
8.01 Holiday Pay		BOXING DAY - IF HOLIDAY ON (IS ADDITIONAL HOLIDAY) THURSDAY, UNION GETS CHOICE OF HAVING TO WORK THURS. OR FRIDAY	—
8.02 Holiday Qualification		BOTH DAYS MUST BE WORKED OR GET SOMEONE TO COVER YOUR SHIFT FOR YOU & TELL FOREMAN	—

48 HR. NOTICE CHANGED TO THE FOLLOWING CONDITIONS

20 MIN WAS REPLACED BY:

Appendix V

Article and Section	Check here if not changed	If different from present contract write the exact new wording below and indicate what is replaced. (Use margin of report when needed.)	Cost (Cents Per Hour)
8.03 Holiday Overtime		2½ HOLIDAY OVERTIME	.22¢/h Imp
9.01 9.02 Vacation Pay		<1 yr ½ day/month 1→3 yr 1 week <3 yr 2% 3→5 yr 2 weeks 3→6 yr 4% 5→10 yr 3 weeks 5→10 yr 6% 10→ yr 4 weeks 10→ yr 8%	2¢/hr/emp.

(margin note: Lowell Has … OVERTIME … CHANGED TO:)

Article and Section	Check here if not changed	If different from the present contract write the exact new wording below and indicate what is replaced. (Use margin if needed.)
10.04 Seniority on Layoff		CHANGE: 'ARE ADEQUATELY QUALIFIED' TO: 'HAVE THE REAL ABILITY' /HAS ADD: IF THE EMPLOYEE IS BUMPED DOWN TO A LOWER PAYING JOB HE WILL RECEIVE 1 WEEK'S PAY AT HIS REGULAR PAY AND THEN PAY AT THE NEW JOB LEVEL
13.01 Wages		Report the increase in wages (in cents per hour). Note: This is the increase which will be given each year of the three-year contract. If there is a strike, report last position. Increase in Hourly Wage Rate: 15% = 75¢ Per year Aver.

Appendix V

14	Benefit	Check here if not changed	Cost to the Company in Cents Per Hour*
Hospital- ization	(a) Life Insurance	✓	
Medical	(b) Hospitalization	✓	
	(c) Operations _INDEMNITY PAYMENTS — CO. PAYS 60%_	✓	
	(d) Doctor Visits CO. PAYS 50%		17.28¢/hr/emp
Pension	(e) Pension SEE BELOW	—	25¢/hr/emp

*Report the changed cost, if any, to the company for each benefit (your last position, of course). Report in terms of the equivalent of cents per hour (an improvement in life insurance, for instance, costing the company, say, $10 a year would be reported as ½ cent per hour, assuming 2,000 working hours in a year).

TOTAL COST
1.32/hr/emp.

DENTAL PLAN — 40% CO. PAYS | NOT EASY TO COST
60% EMPLOYEE PAYS
NO ORTHODENTAL WORK OR FALSE TEETH

PENSION — PROFIT SHARING RETIREMENT PLAN
— EMPLOYEES CAN CONTRIBUTE UP TO $1,000/yr.
THE COMPANY WILL PAY HALF OF WHAT THE EMPLOYEE PUTS IN, IF PROFITS ARE SUCH THAT IT IS POSSIBLE.

Appendix V